IN THE
NATIONAL INTEREST

General Sir John Monash once exhorted a graduating class to 'equip yourself for life, not solely for your own benefit but for the benefit of the whole community'. At the university established in his name, we repeat this statement to our own graduating classes, to acknowledge how important it is that common or public good flows from education.

Universities spread and build on the knowledge they acquire through scholarship in many ways, well beyond the transmission of this learning through education. It is a necessary part of a university's role to debate its findings, not only with other researchers and scholars, but also with the broader community in which it resides.

Publishing for the benefit of society is an important part of a university's commitment to free intellectual inquiry. A university provides civil space for such inquiry by its scholars, as well as for investigations by public intellectuals and expert practitioners.

This series, In the National Interest, embodies Monash University's mission to extend knowledge and encourage informed debate about matters of great significance to Australia's future.

Professor Margaret Gardner AC
President and Vice-Chancellor,
Monash University

IAN LOWE
AUSTRALIA ON THE BRINK: AVOIDING ENVIRONMENTAL RUIN

MONASH
UNIVERSITY
PUBLISHING

Monash University Publishing
Matheson Library Annexe
40 Exhibition Walk
Monash University
Clayton, Victoria 3800, Australia
https://publishing.monash.edu

Monash University Publishing brings to the world publications which advance the best traditions of humane and enlightened thought.

ISBN: 9781922633972 (paperback)
ISBN: 9781922633996 (ebook)

Series: In the National Interest
Editor: Greg Bain
Project manager & copyeditor: Paul Smitz
Designer: Peter Long
Typesetter: Cannon Typesetting
Proofreader: John Mahony
Printed in Australia by Ligare Book Printers

A catalogue record for this book is available from the National Library of Australia.

The paper this book is printed on is in accordance with the standards of the Forest Stewardship Council®. The FSC® promotes environmentally responsible, socially beneficial and economically viable management of the world's forests.

AUSTRALIA ON THE BRINK: AVOIDING ENVIRONMENTAL RUIN

Why do we need to be thinking about the future of our environment? Don't most of us live securely and comfortably in modern Australia? Are there really serious problems we need to be considering? Can't we assume that our lives will continue to improve?

Seventeen years ago I wrote a short book called *A Big Fix*. It was subtitled *Radical Solutions for Australia's Environmental Crisis*.[1] A review began 'What environmental crisis? I just looked out the window and things look fine!' At a basic level, the reviewer had a point. Things do look fine in modern Australia. Most of us live in the cities, and urban air quality is the one major environmental indicator that has actually improved in recent years. We banned the use of tetra-ethyl lead in petrol and this move enabled the introduction of technology which has cleaned up motor vehicle exhausts. The good is gradually being undone as more of us drive large, inefficient four-wheel-drive vehicles longer distances, but urban air quality is actually better than it was thirty years ago. Sadly, this is the one exception—almost all of the other indicators show that our environment is deteriorating.

In 1996, the first independent national report on the state of the environment was produced by an advisory group that I chaired.[2] We found that much of our environment was in good condition by international standards. Some of our approaches were recognised as models of good practice. We also said we had some serious problems that needed to be tackled to achieve our stated goal of living sustainably, not reducing opportunities for future generations. At the top of the list was the loss of our unique biodiversity. The other serious problems we identified were loss or degradation of productive land; the state of our inland rivers, especially the Murray–Darling system; pressures on the coastal zone from increasing population; and the release of increasing amounts of greenhouse gases contributing to global climate change. Five subsequent reports, the most recent irresponsibly kept secret by the Morrison government until after the 2022 election, have shown that those serious problems are all getting worse, and all called with increasing urgency for action to protect our future.

We are clearly not living sustainably, even though our governments all signed up to that aspiration thirty years ago. In December 1992, the Council of Australian Governments adopted the *National Strategy for Ecologically Sustainable Development*.[3] So, at least in principle, our national government and those of our states and territories should be living by the ideals spelled out in that document: meeting our needs in ways that do not reduce opportunities for future generations, striving for equity within and between generations, recognising the global implications of our policies and actions, maintaining the integrity of natural systems, and protecting our biodiversity. As I was preparing to write this book, the Albanese government released its response to Professor Graeme Samuel's 2019 review of our environmental laws.[4] Acknowledging that the laws are not protecting our environment, it said 'Nature is being destroyed' and stated a goal 'to halt decline and repair nature'. Environment Minister Tanya Plibersek promised to introduce in 2023 'stronger laws designed to repair nature, to protect precious plants, animals

and places', setting standards that new developments must meet, and establishing an Environment
Protection Agency (EPA) to enforce those standards.
In a long-overdue pledge to take action, the minister
committed to protecting 30 per cent of our land and
oceans by 2030, to reduce waste, and to build 'an
economy focussed on recycling, reuse and repair' with
the aim of ending the era of species loss.

These are great aspirations, consistent with the
National Strategy. It remains to be seen whether
this government is able to hold the line against the
powerful forces promoting irresponsible economic
development. Those forces have consistently prevailed
since 1992, causing what I called nearly twenty years
ago an environmental crisis. The need for a concerted
response is now critical.

AT THE EDGE OF THE CLIFF

The loss of species is the most important issue we
confront because extinction is permanent; there is
no way of restoring an extinct species. As Australia
has been an island continent for all of geological
time, many of the plants and animals that are found
here do not live anywhere else on the planet. Yet we
have the worst record of mammal extinctions of any

affluent country, and a recent report documented a frightening number of species now at risk.

An April 2018 paper in the journal *Pacific Conservation Biology* reported that we have lost thirty mammal species and twenty-nine bird species since 1788.[5] This has been described as 'the highest mammal extinction rate in the world'. And the damage is continuing. That paper gloomily concluded that seventeen unique Australian birds and mammals are likely to disappear within twenty years. The birds most at risk are located in south-eastern Australia, while the mammals most at risk live in regions of more recent development, particularly the Northern Territory. The bird species that are more likely than not to go extinct in the next two decades include the King Island brown thornbill, the Orange-bellied parrot, the King Island scrub-tit, the Western ground parrot, the Plains wanderer, the Regent honey-eater, the Grey-range thick-billed grass wren and the Herald petrel. Mammals that are more than 25 per cent likely to be extinct on that timescale include the Central rock rat, the Northern hopping mouse, the Carpentarian rock rat, the Christmas Island flying fox, the Black-footed tree rat, Gilbert's potoroo and Leadbeater's possum.

We know what causes species extinction: loss of habitat, introduced species and chemical pollution.

All those pressures are more or less directly due to the demands of the human population. Those forces are now being supplemented by climate change. The three things plants need to grow are warmth, moisture and carbon dioxide. We are changing those drivers of plant growth. As different plants are responding in different ways to the changing circumstances, the balance between plant species is changing. This affects the diet of the herbivores, which in turn changes what is available for predator carnivore species. So we are altering the whole balance of nature, in ways that cannot be predicted with any certainty.

The biodiversity section of the most recent national report on the state of the environment makes very gloomy reading.[6] While some effort has gone into establishing protected areas, excluding predators like feral cats to give native species a fighting chance, nationally the outlook is grim. Habitat loss and degradation are causing 'persistent and sometimes irreversible impacts'. Many ecosystems are experiencing cumulative and compounding pressures that could lead to collapse, characterised by the loss of key defining features and functions. The fundamental problem is that our environmental laws are just not up to the task of protecting our dwindling biodiversity and our threatened ecological communities, as the Albanese government has now recognised. As the title of this essay says, we are on the brink. Urgent concerted action is now needed to avoid disastrous outcomes.

The loss of biodiversity is the most important problem because it is irreversible, but the most urgent task now is to slow climate change As I was writing this, the Kimberley region was largely under water, hit by the worst floods in the history of Western Australia. Settlements along the Murray River in South Australia were desperately building levees to hold back flood-water. The road I recently travelled between Forbes and West Wyalong in New South Wales threaded a narrow path between huge lakes, while the Lismore region was still recovering from three 2021 floods which historically would all have been one-in-a-hundred-year events. The eastern states were devastated by the 2019–20 Black Summer fire season, which destroyed an unprecedented 21 per cent of our bushland. The Great Barrier Reef appears to be in terminal decline from coral bleaching episodes year after year. While the world is now about 1.1 degrees Celsius warmer than it was a hundred years ago, the Australian mainland is on average nearly 1.5 degrees warmer, and large areas in the middle of the continent have warmed by about 2.5 degrees. Many rural areas have been hit by a devastating series of events, causing extreme economic and social distress. The most disturbing statistic in the 2022 *State of the Climate Report* revealed that more than forty days in 2020 were in the hottest 1 per cent ever recorded.[7] The Climate Council estimated the cost of the property damage in the last five years to be about $22 billion, with the 2021 floods in Brisbane alone causing $1.7 billion in property damage and over $250 million in lost agricultural production.[8]

While the recent impacts of climate change are very serious, they are only a foretaste of what is to come. There are very long time lags in the climate system. The most obvious reason for this is the long time that the greenhouse gases we are releasing will remain in the atmosphere. The largest single contributor to global climate change is the increasing amount of carbon dioxide in the air. Tracking the record back through the air bubbles trapped in polar ice has shown that the level of carbon dioxide has varied over the last million years between about 180 and 280 parts per million. The figure now is about 415 parts per million, half as much again as the highest pre-industrial level—a direct result of the burning of huge amounts of fossil fuels: coal, gas and oil. We are seeing the increasing average temperatures that were projected in the 1890s by the Swedish scientist Svante Arrhenius, who warned that burning greater amounts of coal could eventually change the climate.

The carbon dioxide we are putting into the air today will still be there at the end of this century, so climate change will continue for the rest of the century. We cannot stop it. However, the scale and pace of change will be determined by the amount of additional greenhouse gases we release. We are literally creating our climate future by our decisions about which forms of energy to use and how much of them to use.

Recent trends give an indication of the problems we are facing. Catastrophic fire seasons since Europeans arrived here occurred in 1851, 1939, 1983, 2009 and 2020, so in round figures, the intervals between those terrible summers have been ninety years, forty-five years, twenty-five years and eleven years. You don't have to be a mathematical genius to discern the obvious pattern. Asked if the dreadful 2019–20 fires were an indicator of what we might expect in the future, one expert said that by about 2040, it would be a normal year. Similarly, disastrous floods in the Brisbane Valley occurred in 1896, 1974, 2011 and 2022, so intervals of eighty years, forty years and ten years—the same sort of pattern. Wherever we look, we see an accelerating trend of severe events. As the most recent *Summary for Policymakers* produced by the Intergovernmental Panel on Climate Change (IPCC) said:

> The cumulative scientific evidence is unequivocal: climate change is a threat to human well-being and planetary health. Any further delay in concerted anticipatory global action on adaptation and mitigation will miss a brief and rapidly closing window of opportunity to secure a liveable and sustainable future for all.[9]

That is a sobering warning: the window of opportunity to secure a liveable future is 'brief and rapidly closing'. A group of former senior defence figures recently issued a joint statement saying that the greatest threat to our national security is not any conceivable military attack, but climate change, which, they said, now demands a national mobilisation of resources comparable with that which would be required in wartime. It is no longer acceptable to see environmental issues in general, and climate change in particular, as a subsidiary problem to be considered when every other policy question has been resolved. It has to be at the top of the to-do list for all levels of government. Many local authorities have now accepted that we face a climate emergency, but this has not invariably sparked a concerted response.

THE GLOBAL PICTURE

Of course, we need also to think about the global picture; in the unconsciously wise words of a former national minister, 'We can't behave as if Australia was an island.' But the global situation is, if anything, more serious. In 1992, a group of leading scientists from around the world published a warning to humanity, setting out the environmental risks and calling for an urgent response. Basically, there was no discernible response to that warning. More recently, the fifth report in the United Nations Environment Program series on the Global Environmental Outlook said that the recent changes were not just unusual but 'unprecedented in human history'. It went on to warn that several critical local, regional or global thresholds had been exceeded, creating the likelihood of 'abrupt and probably irreversible changes to the life support systems of the planet'. Again, it warned that business-as-usual was no longer an acceptable approach.[10] Yet, unacceptable as it was, most decision-makers continued to behave as if there were no need to respond to the science.

In 2017, the world's scientists issued a second warning, summing up what had happened in the twenty-five years since their 1992 report.[11] This plea for action was endorsed by over 15 000 scientists, myself included. The report found one success story: the warnings about depletion of the ozone layer had led to concerted global action to reduce the release of the chemicals causing the problem. The situation has now stabilised and the ozone layer is expected to recover over the next fifty years. Every other serious problem, we said, is getting measurably worse. Fresh water per capita has been reduced by 25 per cent in the last twenty-five years. The world fish catch has shrunk by 20 per cent. The world has lost 100 million hectares of net forest area. The number of ocean dead zones has almost doubled. The rate of release of greenhouse gases has continued to increase, and climate change is accelerating.

Perhaps most worryingly, species abundance was in 1992 down to 60 per cent of the 1970 level; by 2017 it was only 40 per cent of the 1970 level. This 'mass extinction event', the sixth in the planet's history, means the world is losing species at a similar rate to the past great extinction events, such as the famous one that marked the end of the age of dinosaurs. The 2022 *Living Planet Report* confirmed the worsening trend.[12]

It gave the results of a study of 5200 vertebrate species: mammals, reptiles, birds, fish and amphibians. The average abundance of those species is now 31 per cent of the 1970 figure. On current trends, we could lose as many as a third of all known species this century. That would be a catastrophic loss of biodiversity.

We are pulling random bricks out of the wall of life, but although whole sections are going to collapse, our science is not sufficiently comprehensive to be able to predict the consequences. Since we have only identified and characterised a minority of the other species that we share the world with, we cannot possibly be aware of the full extent of the damage we are doing. It is fair to say that it is almost certainly worse than we currently understand. We don't know what we are losing.

A series of other scientific studies have sounded similar warnings. The Global Footprint Network measures our demands on natural systems. It shows that we are now using about one and a half times what can be sustainably produced, thus inevitably depleting our resource base. The Stockholm Environment Institute has examined nine planetary boundaries and found we are already outside the safe operating space for *four* of the nine. The IPCC has warned that the world has warmed by just over 1 degree since pre-industrial times as a result of the burning of fossil fuels, with consequences that are already serious, but we are on track for an increase of 3–4 degrees if there isn't a much more urgent decarbonisation of our energy systems. Even the World Economic Forum, the big end of town at the global level, said at the end of its 2008 Dubai Summit on the Global Agenda that the observable problems show that the current *economic* system is not sustainable. Further increases in human consumption will obviously worsen the situation. But almost every government in the world aims to grow its economy. It would be morally difficult to justify freezing total global consumption at the current level, since hundreds of millions do not have enough to eat, about a billion don't have clean drinking water, and hundreds of millions lack decent shelter and sanitation. While the number of millionaires and billionaires keeps increasing, huge numbers of people are still living in conditions that are little better than they were in the thirteenth century.

So the global problem appears almost intractable: concern for social justice demands improving the living conditions of the poorest people in the world, which requires making available to them increased quantities of natural resources, but the overall scale of resource use is already much greater than can be sustained by the natural systems of the planet. Summarising the scale of the problem, a 2015 UN report on the economic outlook for the Asia-Pacific region called for 'a new industrial revolution' which would meet the legitimate material expectations of the region's people using 20–25 per cent of the resources per capita that is current practice.[13] It is no overstatement to say that would be a huge challenge.

While it was attacked and not taken seriously at the time, the global problems we are now seeing were projected fifty years ago in the famous first report to a group of like-minded, future-concerned thinkers known as the Club of Rome. *The Limits to Growth* said that if the existing trends of growth in population, food production, industrial output, resource depletion and pollution were to continue, limits would be reached within a hundred years—that is, by 2070—with the most probable result 'a rather sudden and uncontrollable decline in both population and industrial capacity', beginning about 2030. It also found that it would be possible to alter those growth trends and establish conditions of economic and ecological stability that could be sustained far into the future, meeting the material needs of every person on Earth.[14] Because the warnings of the report were not taken seriously, those growth trends have continued. A study by the CSIRO's Dr Graeme Turner compared the 1972 projections with fifty years of data. All those growth trends have continued, putting the global systems on track for 'rather sudden and uncontrollable decline'. What the 1972 report called the 'standard world model' had the global population peaking before 2050, then collapsing to about 60 per cent of the top level. The bio-ecologist William Rees has euphemistically called this 'a major population correction'.

It is a dismal future: the partial or complete collapse of civilisation and the deaths of billions of people. Many reputable analysts now think that downfall is possible; indeed, quite a few think it is a likely outcome of the development path we are now following. There are even some respected thinkers who gloomily see some sort of ruin as almost inevitable. At the 2022–23 Woodford Folk Festival, legendary folk singer Eric Bogle sang a cheerful ditty reflecting on our collective refusal to face up to the global problem, which ended, 'Some will drown, some will starve, some will fry; it's the Armageddon Waltz and we're all going to die.'

I do not accept that gloomy conclusion. I recognise that there is not a single inevitable future, but rather there are many possible futures. Which one eventuates will be the product of our decisions and actions as individuals, members of households and members of communities. We have a wide range of alternatives. But we need to recognise that the ruin projected fifty years ago in the first report to the Club of Rome is now a likely outcome of our collective failure to take that warning seriously. That report asked the obvious question, 'Is the future of the world system to be growth, then collapse into a dismal, depleted existence?' It followed that disturbing thought with the obvious answer, that the 'dismal, depleted existence' is inevitable only if we assume that we will not change.

Technological innovation is helping us to maintain our material living standards using less resources, but the report warned that technological optimism is a dangerous response. 'Faith in technology,' it counselled, 'can divert attention from the most fundamental problem—growth in a finite system— and prevent us from taking effective action to solve it.' To state the obvious, if technical innovation were to halve the resources needed to maintain the present material living standards, we would be back within the sustainable limits of natural systems. But if the global economy were to continue to grow at the present rate, we would be back in the same mess before the middle of the century. As the 1972 report said, unless we recognise that there are limits to growth and develop systems that allow us to meet our needs within those limits, a catastrophic future is inevitable. That is the fundamental issue. It is an urgent matter to slow climate change and halt the loss of our unique biodiversity. But those efforts will be in vain if we remain committed to growth.

As discussed earlier, Australia has an alarming number of species facing extinction and is experiencing accelerating impacts of climate change. We need to develop adaptation strategies for the changes we can't prevent, as well as taking action to protect what remains of our unique biodiversity. Now that we are aware of the potentially catastrophic consequences of our historic approach to development, we have a responsibility to change.

INVESTING IN CLEANER ENERGY

The first and most obvious point to make is that we cannot solve the global problems by ourselves. In the urgent case of climate change, the greenhouse gas emissions resulting from Australia's energy use and land clearing are only about 1.2 per cent of the global figure. We are actually responsible for much more than that, because we export coal and gas which is burned in the importing countries, mostly China, Japan and South Korea. However, when we add in the emissions resulting from our fossil-fuel exports, we are still only responsible for about 4 per cent of the global total. We have no control over the other 96 per cent of global emissions.

Nonetheless, we should recognise that there are very few countries that emit a larger fraction of the global total than we do, if we accept responsibility for our fossil-fuel exports. China accounts for about 24 per cent of the world's emissions, the United States about 12 per cent, India 6.6 per cent, Russia 4.8 per cent and Indonesia 4.5 per cent, but apart from these five nations, no other country is responsible for more than we are. Our total emissions are one and a half times those of Japan, two and a half times those of Germany or Canada, and more than three times those of South Korea. Our annual emissions of green-house gases add up to about 530 million tonnes, or about 21 tonnes per person. So on average we are each releasing about 60 kilograms a day of carbon dioxide, as if we were each burning about 15 kilograms of coal every day of the year. That is a direct consequence of our energy use. If you divide Australia's total energy use by our population, it is equivalent to every one of us using about 6 kilowatts continuously. You are not conscious of using energy at anything like that rate as you sit quietly reading this book. It is a consequence of the fact that energy is used in every aspect of your life: the buildings you use, the clothes you wear, the food you eat, your entertainment, your transport, heating and lighting. We are so dependent on that energy that whenever there is an interruption to its supply, chaos results.

We are a major contributor to the emissions problem, so we certainly have an obligation to do our share of the global effort to slow climate change. Under successive Coalition prime ministers—John Howard, Tony Abbott, Malcolm Turnbull and Scott Morrison—we acquired a global reputation for obstructing progress and using blatantly dishonest accounting tricks to cover up our inaction. But after a decade of procrastination, we now have a national government that has a target of net zero emissions by 2050, while several state and territory administrations have more ambitious goals. It is fair to say, however, that we don't yet have in place policies and programs that will achieve those targets.

The former Australian chief scientist Dr Alan Finkel wrote a 2021 *Quarterly Essay* on this topic, 'Getting to Zero'.[15] He pointed out that achieving the target of zero emissions will be 'difficult, but not impossible'. Because our civilised world is critically dependent on the use of massive amounts of energy, we can't simply stop using fossil fuels. We need to replace them with low-carbon energy sources, principally solar, wind and hydro. Because the sun doesn't shine at night and the wind doesn't always blow, transforming our electricity system to use the natural energy flows from the sun and the wind demands a huge investment in storage systems. Hydro-electricity relies on a form of storage we have used for a very long time: water in reservoirs that can be released when we need to use it to generate electricity. We have become more aware of the ecological damage of large dams over the years, but there are still additional storage schemes under construction despite those issues, such as Snowy 2.0. More environmentally responsible than large impoundments are small-scale pumped hydro schemes. A study by Australian National University (ANU) scientists identified several thousand potential sites around the so-called national grid, the electricity system of the eastern states and South Australia, of which only about fifty would need to be built to provide enough storage for the system to run totally on the renewable sources of solar and wind.[16]

The NSW Coalition government has called for commercial tenders to build eight such schemes as a crucial part of its plan to receive 95 per cent of its power from solar and wind by 2030. South Australia, meanwhile, has invested in very large battery storage systems. While these initiatives of the Weatherill ALP government were criticised by the national Turnbull regime, more recent SA governments of both political persuasions have doubled down on the approach because it makes sense. Over 2022, South Australia got nearly 70 per cent of its electricity from solar and wind, using its battery storage to even out the intermittent nature of the natural energy flows. That state has a realistic chance of meeting all its power needs from solar and wind by 2030. It already achieves that for significant periods, including a run of ten days in December 2022 when solar and wind actually generated more power than the state used, with the surplus exported to Victoria.

In overall terms, phasing out fossil fuels from the electricity system has been made easy and cost-effective by advances in the technology providing huge improvements in the economics. In my recent book about Australia's role in the nuclear industry, I quoted the average global prices of power from the various major sources.[17] In 2010, electricity from gas typically cost about 8 cents per kilowatt hour (kWh), coal about 11 cents, nuclear 12 cents, wind 14 cents and solar 35 cents. By 2020, the average price of solar power was 3.7 cents per kWh, wind 4.1 cents, gas and coal still about 8 cents and 11 cents respectively, and nuclear about 16 cents. While it was true a decade ago that the clean-energy technologies carried a cost penalty, today they are by far the cheapest power systems. In the Northern Hemisphere, the cost advantage is so obvious that old coal and nuclear power stations which long ago amortised their capital costs are being closed down, because just the running costs make their power more expensive than can be obtained from solar farms and large wind turbines. The International Energy Agency (IEA) figures for the changes in delivered electricity between 2019 and 2020 are startling: nuclear about 100 terawatt hours (TWh) less, gas 130 TWh less, coal 500 TWh less, and renewables about 400 TWh more. The decline in total global power use in 2020 represents the impacts of the COVID-19 pandemic.

The investment trend is quite remarkable too. In 2019, the world installed about 70 gigawatts (GW) of new fossil-fuel generators and about 170 GW of new renewables. In 2020, about 190 GW of new renewables were installed—107 GW of solar, 65 GW wind and 18 GW hydro—but only about 40 GW of new fossil-fuel power, all of it gas-fired. There was actually less coal-fired power at the end of the year than there had been at the start, as closures slightly exceeded new capacity for the first year in living memory.

There is still a small investment in nuclear power, nearly all of it in China, but the three nuclear power stations currently being built in Western Europe are all years behind schedule and billions of euros over budget. In 2020, according to the IEA, about 8 GW of new nuclear power capacity came online but 5 GW of existing power stations were decommissioned, giving a net gain of 3 GW compared with 190 GW of renewables. It is very clear which way the world is going.

When Professor Barry Brook and I joined forces in 2010 to write a book giving the arguments for and against nuclear power for Australia,[18] there was still a credible case that nuclear could be the cheaper low-carbon power source. While a group of pro-nuclear zealots is still promoting the so-called small modular reactors, the economics of these appear to be beyond the reach of even the most creative arithmetic. It remains true, as the Uranium Mining, Processing and Nuclear Energy Review commissioned by the

Howard government found in 2006, that it would require significant public subsidies to make nuclear power look cost-effective. That report also concluded that it would take at least ten years and more probably fifteen to build one nuclear power station in Australia, given that we do not have experience of constructing large-scale nuclear reactors or regulating the technology. That seems realistic; a recent BBC documentary about the first nuclear power station to be built in the United Kingdom for decades, Hinkley Point C, estimated it would take at least another ten years to complete the project, which is already well underway. Even if there were the political will and social approval for nuclear power, it is far too slow and far too expensive to be a realistic response to the urgent need to phase out coal-fired power. Going nuclear would also require overlooking the problems of needing to manage radioactive waste for all geological time and needing military-scale security to prevent the misuse of fissile material.

Taking all those factors into account, nuclear power just doesn't make sense for Australia. There was a flurry of excitement in the media recently when a nuclear fusion experiment briefly produced more energy than had been needed to start the reaction. While there is the dream we might one day have unlimited energy from fusion of light elements like hydrogen, there is no chance of that being realised quickly enough to help slow climate change. The standard physicists' joke, which I heard as an undergraduate, was that commercial fusion power is fifty years away, and probably always will be. The technical problems of harnessing fusion are truly formidable.

USING ENERGY EFFICIENTLY

If we are serious about slowing climate change, the quickest and most cost-effective way to reduce our emissions will be to increase the efficiency of using energy. Nearly fifty years ago, the US energy analyst Amory Lovins made the perceptive observation that 'people don't want energy, they want hot showers and cold beer'. We live more comfortably than any previous generation because fuel energy allows us to easily do tasks that were previously either impossible—air travel over long distances, rapid access to information, keeping beer cold—or required real physical effort—digging holes, washing clothes, kneading dough. But it is those tasks that we want to have done rather than needing specific amounts of energy; I have never heard a householder say they need more megajoules. Much of the technology we use is still very inefficient. To quantify what could be achieved, a report prepared for the Howard government in 2003, *Towards a National Framework for Energy Efficiency*, estimated that Australia's emissions could be reduced by 30 per cent simply by using cost-effective existing technology.[19] The report used the idea of payback time: how long the cost of a new device will take to be recovered by using less energy and thus saving money. The criterion the report writers used was four years. About one-third of our emissions could be eliminated by appliances that would save enough money to recoup their capital cost in less than four years. With the improvements in technology since then, the figure would now be even larger. It is a scandal that we allow the sale in Australia of many appliances that could not legally be sold in the European Union. Some could not even be sold in the more progressive states of the USA.

A 2018 international study found that our efficiency standards and policies were the worst in the entire Organisation for Economic Co-operation and Development.[20] We are one of the only advanced countries that does not have vehicle efficiency standards. In fact, the average fuel efficiency of the Australian vehicle fleet has scarcely improved in the last fifty years. All of the technical improvements in vehicle design, engines, drive trains and tyres have been countered by vehicles getting steadily larger and more likely to have fuel-hungry features: automatic transmission, power steering, complex entertainment systems, four-wheel drive to cope with the difficult terrain of suburban streets.

Our building standards are similarly appalling, meaning that much more energy is required to keep our dwellings and offices comfortable in summer and winter than would be the case if they were well insulated and sensibly oriented. When I returned to Australia from the United Kingdom in 1980 and was trying to find somewhere to live, I noticed that many houses lacked a range of useful design features: they had large windows facing west to catch solar energy on hot summer afternoons, no eaves to protect from undesirable solar heating, few windows on the north side to allow free warming on winter days, a dark roof to absorb as much solar energy as possible. I found a place that was well designed and properly oriented to get free solar heating in winter and good cross-breezes in summer; indeed, I have moved three times in the forty-three years I have lived in Queensland and have always been able to find similar places to live. But while in 1980 about 5 per cent of dwellings in south-eastern Queensland were air-conditioned, recent data reveal that about two-thirds are now cooled in this way. This is a reflection of the fact that many houses and flats are poorly designed and unintelligently oriented, sentencing their occupants to paying large amounts of money to be comfortable. We would use less electricity to maintain our material living standards if our buildings and our appliances were required to meet modern standards of efficiency.

Now that solar panels produce the cheapest electricity, power at a much lower cost than the average price of electricity from the grid, it would make financial sense to insist that buildings which are mostly occupied during the day should have solar panels on the roof. Schools, university buildings, offices and shopping centres should as a matter of course generate most or all of their power needs from rooftop solar panels, as well as possibly selling any surplus that they don't use.

One of the worst wastages of energy is electric water heating. Most of our electricity still comes from fossil fuels, so the process involves burning coal or gas to boil water and produce steam which drives a turbine, turning the heat energy inefficiently into electricity which we distribute through wires, losing energy all the way to our dwellings, where we turn the electricity into lighting or heat or movement of some sort in a mixer or a fan. Solar water heating makes economic sense almost everywhere in mainland Australia. When I and my Griffith University colleagues conducted a survey nearly forty years ago of users of solar hot water, we found that the average time for the savings in power to repay the capital cost of the solar device varied from about four years in Queensland to about six in most other parts of the country. Only in Tasmania and in Victoria south of the Great Dividing Range was the payback time comparable with the expected life of the product. I was so impressed with the data that I immediately bought a solar water heater. When I moved house ten years later, I bought a newer and better solar water heater. Where shading or the fragility of a roof rules out solar hot water, heat pumps dramatically improve the efficiency of producing hot water, thus reducing power demand (and, of course, power bills). The electric water heater should be now only found in folk museums, showing modern people how wasteful we were in the past.

The recent revolution in lighting has also brought huge savings. The incandescent bulbs that were the universal technology when I was young did literally produce more heat than light. They gradually gave way to fluorescent tubes, then compact fluorescent bulbs and more recently light-emitting diodes (LEDs). These use about one-tenth of the electricity to produce the same amount of light as the old light bulbs. They also produce other savings. I was told by a traffic engineer that the bulbs in traffic lights used to be replaced when they had only been in operation for about half their expected life, because it is a safety issue—there has to be a red light to stop vehicles. Replacing the single bulb with a mosaic of LEDs has dramatically reduced maintenance costs because it doesn't matter if 10 per cent or even 20 per cent of the diodes fail; there is still a traffic signal. Similarly, about half the air-conditioning load in a typical office building used to be for removing the waste heat from lighting. Now that lights are much more efficient, the air-conditioning does not need to work as hard to maintain a comfortable temperature.

The general point is that we should not assume that electricity demand must keep increasing. Paying attention to the task can reduce electricity demand by turning the power more efficiently into the services we want, or by using natural energy flows such as solar to provide light or heating. As Dr Finkel wrote, 'If our use of energy were more efficient, we would not have to produce nearly as much.' Since there is no energy supply system that is totally environmentally benign, no source of zero-carbon electricity, the highest priority should be to convert energy more efficiently into the services we want.

MODERNISING TRANSPORTATION

In many ways, the easiest part of the need to decarbonise our energy supply is the electricity system, because renewable technologies are now far cheaper than fossil fuel systems. The problem is that power supply only accounts for about 170 million tonnes of the carbon dioxide emitted by Australia. We need also to address the other 360 million tonnes. As Dr Finkel put it, 'we will need to change the way we farm food and process it, our vehicles and transport systems, our building designs ... heating and cooling systems, our industrial processes'. All of these areas are much more challenging than the comparatively straightforward process of closing down fossil-fuel power stations and upgrading the grid to make more use of renewables.

The second-largest contributor to our greenhouse gas emissions is transport. Our transport systems are almost entirely powered by fossil fuels, mostly liquid fuels derived from oil. Hybrid vehicles have been on the market for quite a few years. These typically use about half as much fuel as the standard internal combustion engine burning petrol or diesel fuel. The extra capital cost has meant that they have appealed mainly to those who drive long distances and can realistically recoup in lower fuel costs the higher price, such as drivers of taxis or Ubers. Our buses mainly use diesel, but an increasing number are gas-powered. When I was on Brisbane City Council's environment advisory committee, we persuaded the council to buy some gas-burning buses to improve urban air quality, since diesel engines are notorious emitters of fine particulates that contribute to respiratory distress. Having bought the new buses, the council found they were actually cheaper to run, so they embarked on a steady process of replacing the old diesel buses when they were retired by newer models using gas. On a recent trip to Brisbane, I noticed that there is now also a small group of electric buses. The Queensland Government has decided that we should move systematically to using electricity for buses.

The question of whether electric vehicles help to slow climate change is complicated by the question of where the electricity comes from. When we get most or all of our power from renewables, it will clearly be beneficial to use electric vehicles. At the time of writing, however, about 70 per cent of grid electricity was still coming from burning coal. The process of turning the heat energy of coal into electricity is very inefficient, but that electricity is turned very efficiently into motive power by electric engines. The process of turning the heat energy of petroleum fuels into motive power by an internal combustion engine or diesel engine is also very inefficient, but the petroleum fuels are not as carbon-intensive as coal. Much of the energy comes from turning hydrogen into water, rather than from turning carbon into carbon dioxide. As you can see, it is a very complicated calculation. Still, the scientists whose work I respect mostly conclude that electric vehicles are better for the climate, even with the present balance of where the electricity comes from. That balance is continually shifting as every year more renewables replace coal and gas, so the climate advantage of electric vehicles will continue to improve.

The current biggest obstacle is the extra cost, putting electric cars out of the reach of people on average incomes. Those with inside knowledge of the vehicle industry believe that the price of new electric cars will fall rapidly. They expect the point at which the overall cost of electric cars will be competitive to come later this decade, given that the running cost is typically about one-third of what is required for petrol or diesel vehicles. Some major car companies have already announced dates beyond which they will no longer build vehicles requiring petroleum fuels, a move driven in part by European nations legislating a phasing-out of those vehicles over the next decade. It is clear the ways of the world of transport are changing. The United Kingdom has legislated that no new petrol or diesel vehicles can be registered after 2030 and no new hybrids after 2035. Other European countries have legislated an earlier date for the transition, while China has a similar target, and the majority of all new vehicles in Norway are now electric, with a goal of all new vehicles being electric by 2025. Even the US-based General Motors, famous for huge petrol-powered cars, has announced it will only produce zero-emission vehicles after 2035.

The historic obstacle to using electric cars was the problem of storage. I used to show my students a quotation that said the electric car was clearly the way of the future, being clean, quiet, efficient and non-polluting. All it required was an advance on the lead-acid battery for storing the electricity it used. The quotation was dated 1903! For over a hundred years, the weight of lead-acid batteries made the electric car impractical. Now, modern battery technology has changed the world.

There still remains the question of efficiency. I have often speculated what would happen if engineering students were asked to design a transport vehicle to move around a fragile payload, typically weighing between 50 and 100 kilograms. If they came up with a vehicle weighing one and a half tonnes, somewhere between ten and thirty times the load it would carry, I think they would be advised to consider career options for which numeracy was less important. Because the car began life as the horseless carriage, with comfort more important than fuel efficiency, we have a design that is ludicrously inefficient. It is conceivable to have small cars for urban commuting, possibly weighing less than half a tonne. The problem then is one of safety. A light vehicle sharing the road with heavier ones will always come off second best if there is a collision. Because we allow on our roads trucks weighing several tonnes, and large two-tonne four-wheel-drive cars that a colleague calls urban assault vehicles, a commuter would rightly feel unsafe in a small city car. For historic reasons, we provide tax concessions to four-wheel-drive vehicles and huge public subsidies for road freight vehicles. Road user charges that reflect the cost imposed on the community, or carbon pricing based on greenhouse gas emissions, would change the economics and provide real incentives for more efficient vehicles.

Dr Finkel's *Quarterly Essay* speculated that electric cars will be the standard for personal transport within a decade, but he concluded that the scale of batteries required would rule out that approach for long-distance heavy-freight movement, making it more likely that hydrogen fuel cells will be used. The technology is solidly proven. Perth was one of eleven cities round the world that participated in a trial of hydrogen fuel cell buses between 2001 and 2004. I rode in one of the buses and talked to the driver; from the viewpoint of driver or passenger, there is no obvious sign that you are being propelled by hydrogen rather than diesel, gas or electricity. The only visible difference was that the bus was about 30 centimetres higher than normal, to allow for a very large tank to hold enough hydrogen for a day's driving. The process of turning renewable electricity into hydrogen and the hydrogen back into motive power is relatively inefficient, so the cost makes it impractical for a private car. But as road freight receives huge public subsidies, efficiency is probably a less significant factor. In terms of public transport, Auckland introduced hydrogen buses on one route in 2021, and the following year, the NSW Government announced a comparative trial between hydrogen and electric buses.

Dr Finkel also argues that it is impractical to shift 150 000 tonnes of iron ore in a ship powered by batteries, but it could be done with hydrogen fuel cells. I have speculated that we could even see a return of the sailing ship, using wind power for long-distance ocean freight. After all, if there is a steady stream of ships carrying iron ore from Western Australia to China or Japan or South Korea, the speed of the voyage would not appear to be particularly important. It might be cost-effective to take four to six weeks driven by wind rather than two weeks burning hydrogen. Wind power certainly would be better for the climate than using electricity to split water into oxygen and the hydrogen needed to power ocean freight vessels.

It is, again, a common mistake to assume that the transport task is fixed and immutable. In a city, it is a product of decisions about planning and the provision of services. When I was in Sydney in the 1960s, almost everyone lived within walking distance of a railway station, a bus stop or a tram stop. Most people used public transport to commute to work, travel to shops or reach a place of study. Then the widespread availability of cars made it possible to build new suburbs outside the sphere of public transport. Australian cities are now divided into two distinct sections. The inner suburbs are quite densely settled, with many of the services people use regularly within walking or cycling distance, and good public transport services for longer trips. On the peri-urban fringe, nothing is within walking distance, cycling is unsafe and public transport is almost non-existent, forcing those who live there into long car trips. In their book about the problem of car-dependent cities, Professor Peter Newman and Dr Jeff Kenworthy contrasted two new cities built in the 1960s: Almere in the Netherlands and Milton Keynes in the United Kingdom, where I lived when working for the UK Open University in the 1970s.[21] The Dutch city has, on average, twice as many dwellings per hectare as the English one. As a result, a much larger percentage of the urban journeys are less than 3 kilometres: 85 per cent in Almere, 45 per cent in Milton Keynes. This is directly responsible for a startling difference in the mode of transport used. In Milton Keynes, about 60 per cent of journeys are made by car and about 5 per cent on bicycles. In Almere, nearly 30 per cent of trips are made on bicycles and only 35 per cent by car.

People are rational. If our urban areas are planned so that the services we use every day are within walking or cycling distance, we will walk or cycle. If our urban areas are perversely designed so that longer trips are required, we will drive cars. When Maroochydore Council consulted the community about the future of its part of the Sunshine Coast, it was clear what people wanted: compact local urban villages with services within walking distance, connected by good transport systems and allowing protection of the remaining native vegetation to preserve habitat and biodiversity. The council was then forcibly amalgamated with two others to form the Sunshine Coast Regional Council and the vision of *Maroochy 2025* disappeared.

The area of transport that looks most intractable is air travel. As a person who was once a so-called 'frequent flyer', I am embarrassed that my carbon footprint was then dominated by my flights around this vast country. Globally, aviation is responsible for about 2 per cent of greenhouse gas emissions and, unlike the other areas of transport, there is no obvious technical solution that would allow the present level of flying to continue. One of the few positive outcomes of the COVID-19 pandemic is that we have seen dramatic improvements in videoconferencing technology. I now routinely use the new systems to give presentations to audiences in other states, removing the need that previously existed to travel.

Airbus, which has a stated goal of zero-emissions flying by 2035, has joined a New Zealand consortium hoping to develop commercial aircraft powered by 'green hydrogen' using solar energy. This is a brave goal, but I think it is likely that a zero-emissions world will involve much less flying interstate or overseas. Young people in the future will probably read with wonder and disbelief about the short epoch of large-scale international air travel. They will likely be equally incredulous about the burning of transport fuels for gratuitous entertainment. I can imagine a child saying with wonder to their parents, 'Is it true that people used to put fuel in cars and drive them round in circles, just to see which one could go the fastest? Didn't they know about climate change?'

TACKLING OTHER EMISSIONS

We use large amounts of natural gas for heating and cooking, producing about 80 million tonnes of carbon dioxide a year: the third-biggest contribution to our carbon footprint after electricity and transport. In his book *The Big Switch*, Dr Saul Griffith argues these applications should use electricity from renewable sources.[22] He makes the point that heating a saucepan of water using gas is very inefficient, as more of the heat escapes into the kitchen than actually increases the temperature of the water. Burning gas in an enclosed space like your kitchen also pollutes the air and, according to a recent study by the Climate Council, has impacts on respiratory health comparable to having a smoker in the house. Modern electric induction cooktops are much more efficient than old-fashioned electric hot plates or gas burners. Dr Griffith calculates that using an electric induction cooktop powered by rooftop solar panels reduces the cost of cooking to about one-sixth of that required to use gas, at 2022 prices. As discussed earlier, an electric heat pump such as a reverse-cycle air conditioner in our climate typically produces three to four units of useful heat for every unit of electricity driving it, so it is a much more effective way of warming a room in winter or providing hot water than burning gas. It is also important that it can use renewable sources for the electricity, thus removing the present release of carbon dioxide.

Agriculture generates the next biggest chunk of our greenhouse gas emissions. While most people immediately think of the diesel fuel used by tractors and other farm machinery, the largest component of our rural emissions is the burping and farting of methane by ruminant animals: cattle and sheep.

CSIRO has been working to develop a feed supplement from seaweed, with trials showing this can dramatically reduce the amount of methane released by cattle. Comparative health studies suggest we would have lower rates of colon cancer if we ate less red meat, but this dietary change would have comparatively little impact on the scale of our grazing industry because about 70 per cent of the beef and lamb we produce is exported.

Agriculture and mining raise the important broader question of why we need to dig up so many minerals and work our rural land as hard as we do. An economist would say that we need to export those raw materials to enable us to pay for our imports, things we have collectively decided we are not clever enough to make, like T-shirts and runners. (I will return to this issue later, as it poses the basic question of what sort of country we want to be.) When I was young, we manufactured cars and aircraft, various kinds of communications equipment, radio and television receivers, other domestic appliances like toasters and food-mixers, clothing and footwear. In the 1980s, the Hawke government told us we should try to become the clever country. Instead, we have become the stupid country with the trade pattern of a poor developing nation, exporting minerals and agricultural produce to allow us to import manufactured goods. Australia now has the smallest manufacturing sector in the entire developed world, as a percentage of our overall economy. That makes us critically dependent on the world trade system, as we discovered to our discomfort during the COVID-19 pandemic when that system was significantly disrupted.

An economist would probably argue that the world trade system allows us not just to convert iron ore into manufactured steel products, but also to convert iron ore into clothing, or to transform wine and seafood into computers and cars. That is a defensible approach while the system is operating smoothly, but if the first report to the Club of Rome is correct and we are likely to see greater disruption of global systems, we will be dangerously exposed. I believe it would be prudent to work towards being more self-sufficient. This would mean we could reduce the numbers of grazing animals and thus proportionately reduce their emissions. Agricultural machinery uses enormous quantities of diesel fuel, but the great thing about the diesel engine is that it can operate on a wide range of fuels, so we should encourage farms to produce their own bio-diesel from a range of crops: peanuts, sunflowers, canola, macadamia nuts and so on.

The remaining significant sectors contributing to our emissions are industry, mining and buildings. I have already discussed the need for our buildings to be designed and oriented to minimise the need for fuel energy, to maximise use of natural light and flows of energy. Our mining industry is presently dominated by coal, gas and iron ore. The world is going to progressively use less coal and gas, so even if our governments irresponsibly continue to encourage the export of those products, their markets will inevitably decline. But as the world moves to using renewable energy, other minerals will be in demand. As Geoscience Australia points out, we have an amazing proven abundance of the minerals the world will need. We have the largest share in the world of zircon, tantalum, titanium, lead, iron ore, zinc, nickel and gold, as well as the second-largest share of bauxite, cobalt, copper, lithium, tungsten and vanadium. We also have the world's largest share of uranium, still used by some countries for nuclear energy. While it presents other problems, such as the need to manage radioactive waste, it is a relatively low-carbon form of energy. It seems highly likely that our mining industry will expand, moving away from fossil fuels and producing more of the critical minerals needed to harness renewable energy. Decarbonising the mining industry will mean moving away from petroleum fuels and grid electricity to solar and wind power. As mines tend to be in remote parts of the country, it makes sense to use local solar or wind energy with battery storage, rather than burning diesel fuel to produce the power they need.

The biggest users of energy in the industry sector are the large-scale producers of steel, alumina and aluminium. The aluminium industry is notorious for enjoying huge public subsidies and employing relatively few people. The Australia Institute, a Canberra-based progressive think tank, estimated that the public subsidy is so large and the number of people employed so small that it would cost the taxpayer less if the industry were closed down and every employee paid $100 000 a year to go fishing![23] While that is probably not a practical approach politically, it makes no sense to provide huge subsidies for the use of fossil-fuel electricity to allow the local industry to compete with Canadian hydro-electricity or Icelandic geothermal power, both of which are low-carbon sources. There is great interest around the world in so-called 'green aluminium' and 'green steel', recognising that these essential building materials will continue to be needed, so we must develop ways of producing them without the high carbon price now being paid. There has been serious research into the possibility of revolutionising steel-making, using renewable electricity and clean hydrogen rather than coal, but that approach is still at a developmental stage.

Since some of the greenhouse gas emissions will be very difficult to eliminate, reaching net zero will require some measures to take carbon dioxide out of the air. For decades, the fossil fuel industries and their political supporters have been talking about carbon capture and storage as a sort of get-out-of-jail-free card, suggesting that we could continue recklessly burning coal and gas because the emissions would be captured and stored. In principle, it is possible to capture the carbon dioxide from burning coal and gas, then liquefy it and try to inject it into geological strata far beneath the surface. However, there are two major problems with this: first, it would cost a lot of money and significantly reduce the energy output of a power station if the emissions were captured and liquefied; second, there would need to be suitable geological strata near the site. There may be niche applications where there happens to be the right geology near a coal-fired power station, but there is no prospect of scaling the process up to deal with all the carbon dioxide that comes from burning fossil fuels. A quick calculation suggests that if all the carbon dioxide produced by the world's coal-fired power stations could be captured and liquefied, we would be dealing with a volume of liquid comparable with the entire oil industry. This is clearly not a practical solution. Although successive governments in thrall to the fossil-fuel industries have poured huge sums into research and development, so far all that has been successfully captured and stored is taxpayers' money.

There is one form of carbon capture and storage that we know can work: planting trees. We should restore some of the tree cover that has been destroyed by clearing to enable expansion of grazing and agriculture, simultaneously setting up the system to capture and store carbon while reducing the emissions from ruminant animals and cultivation. This would also help protect our remaining biodiversity by providing habitat for native species. In principle, the Australian Government's Emissions Reduction Fund was set up to pay landholders to revegetate their properties. Dr Finkel reported in his 2021 *Quarterly Essay* that the fund has signed contracts with landholders to remove some 200 million tonnes of carbon dioxide from the atmosphere. There has been some criticism that farmers are being paid just to do what they were always going to do, but another former chief scientist, Professor Ian Chubb, recently gave the scheme his tick of approval. It is clearly not a silver bullet, but at least it can take a modest step in the right direction, unlike the proposals to capture and store emissions from power stations.

USING THE WHOLE TOOLKIT

There is an old saying that if you only have a hammer, every problem looks like a nail. We all tend to see the world through the lenses of our values and experience. As a result, we are likely to try to tackle new problems with the tools or approaches we have used before and are comfortable with. Think of a problem like road congestion. Engineers tend to think of technical solutions such as building new roads, widening the existing carriageway or replacing traffic lights with roundabouts. Lawyers tend to think of regulatory responses such as requiring cars to have multiple occupancy or specifying which vehicles can use the road at peak times. Economists tend to think of financial signals such as road-user charges or fuel taxes. Educators tend to think of ways to provide the road users with better information so they can plan their journeys to avoid the bottlenecks. In practice, where we have successfully achieved major changes in human behaviour, we have used all the tools in the toolkit as appropriate. Think of the changes in the last fifty years in our attitudes to smoking in shared spaces or driving while affected by alcohol. In those cases we have regulated, we have educated the community, we have introduced financial incentives or disincentives, and we have deployed new technology where appropriate. If we are to achieve the dramatic change needed in our approach to climate, cutting Australia's domestic emissions from about 530 million tonnes a year to net zero in just twenty-seven years, we need to take a similarly comprehensive approach. As mentioned earlier, a group of defence chiefs said we should put the nation on a wartime footing to meet the challenge. That may be too big a step for timid politicians, but we should certainly be using all the factors at our disposal.

There are limits to what price signals can achieve at the individual level. I can't imagine a person hovering by a light switch as the sky darkens, weighing up whether the few cents it will cost to have the light on is a sensible use of their money. I can imagine a person debating whether to turn on a major power-using appliance like an air-conditioner. We know that the Gillard government's modest carbon price influenced the decisions of large power corporations. Emissions from electricity generation declined while the carbon price was being charged. When Tony Abbott was elected, having promised to remove the charge and then actually doing that, emissions went back up. There is widespread agreement along the political spectrum, from fervent greenies to far-right economists, that a carbon price will encourage the large emitters to curb their release of greenhouse gases. The problem is political. The Liberal and National parties, while generally in thrall to the discredited economic theories of the neoclassical school, are ideologically opposed to a price on carbon. The brutal attack on the Gillard minority government for its modest carbon price seems to have frightened the Australian Labor Party (ALP) into submission, worried that it will be accused of bringing in new taxes.

Of course, government has to use taxation to obtain the revenue for its expenditure, and it is a good principle to raise taxes on behaviour we would like to discourage. Successive governments of both political colours have done this with tobacco, steadily increasing the price of cigarettes to discourage smoking. Releasing carbon dioxide and other greenhouse gases into the atmosphere is as bad for planetary health and the welfare of the community as cigarette smoke is for the lungs of an individual; arguably worse, because the statistics show that about half of smokers have their lives shortened by tobacco, while all of us are affected by accelerating climate change. At some point, we have to overcome the timidity of the ALP and the ideological rigidity of the Coalition and discourage emissions by a price signal. I recently heard an ALP figure say we don't need a carbon price because renewable electricity is now cheaper than coal-fired or gas-fired power. That is true; hard-nosed accountants without a green bone in their body are urging the generators to phase out fossil fuels and expand their investment in solar, wind and storage. The market is working to reduce emissions from the electricity system. However, a carbon price would provide incentives to reduce the other two-thirds of our emissions.

Regulatory measures could rapidly produce major reductions in our emissions. Building standards, appliance efficiency rules and vehicle efficiency standards all have the capacity to make very large inroads, given how far behind world's best practice we are. There is, of course, a time lag, which is why action is urgent. Buildings usually last for decades, so we should be prepared to retrofit the existing stock to modern standards as well as setting those requirements for new construction. Cars don't last as long on average, but we are still talking about ten years or more. That is why several countries have recognised that a 2050 target of net zero emissions means not allowing new models burning traditional fuels after 2035. Domestic appliances and items of industrial equipment typically do not last as long, so the savings that can be achieved by requiring modern standards come much faster. The low-hanging fruit is the removal from the market of the clunky inefficient appliances that are not allowed in Europe but are being dumped on the Australian market. Getting rid of those will save consumers money as well as reduce our carbon dioxide emissions. Of course, restoring our capacity for manufacturing will enable us to produce appliances locally, rather than burning fuel to import them from the Northern Hemisphere. So a determined effort to become a clever country will have environmental benefits as well as the obvious economic and social advantages that would flow from producing the goods we need here—designed for local conditions, using local materials and local skills.

I think it is clear that halting the loss of biodiversity will require regulatory actions. The primary cause of species loss is destruction of habitat. So, as our federal environment minister said in responding to the Samuel review of environmental laws, we need a commitment to increasing protected areas as well as a national EPA with the resources to enforce stronger environmental standards. Excluding feral predators gives native species a better chance of surviving, so we need serious efforts to eliminate introduced predators from our natural areas: our rangelands, our forests, our bushland, our savannahs, our beaches and estuaries. We also need to halt the destruction of natural areas for a wide range of activities: mining, agriculture, urban expansion, offices and shopping centres, road widening. Historically, community groups, volunteer organisations and determined individuals have tried to protect our natural areas from irresponsible development, but environment courts have tended to assume that destructive activities can go ahead unless it can be proven that the impacts would be unacceptable. It is time to reverse the burden of proof. Given the urgency of preserving what remains of our natural areas, those proposing new mines, agricultural expansion or other measures that may damage or destroy those areas should have to prove that the impacts will be acceptable. As well as properly funding Environmental Defenders Offices, as the Albanese government has done, we should also ensure that environment courts are resourced to enable independent scientific assessment of competing claims. Protecting our remaining biodiversity requires our decisions to be based on solid science.

I am tempted to say we don't really need much education to persuade the community of the need to change. The voting pattern in the May 2022 federal election plainly showed that the people are ahead of the politicians, prepared to go against decades of entrenched practice to elect 'teal' independents and Green candidates in what had been regarded safe seats for the Coalition or the ALP. There is a widespread understanding that we need to change. However, I think it is still true that most people 'underestimate the extent to which society must change in order to be sustainable and the significant limits that we'll need to place on our consumption', as Dr Aaron Karp wrote in response to a discussion paper I released in 2022. The arithmetic is simple. Total human consumption is now about 150 per cent of what natural systems can sustainably produce. That means getting back into balance requires significantly reducing the overall demands of the human population. But it would be unacceptable to condemn those now living below the poverty line, without clean water or sanitation or adequate nutritious food, to remain in that deprived situation. The inescapable conclusion is that those of us living in comparative comfort in countries like Australia and New Zealand, as well as regions like North America and Western Europe, will need to reduce our resource demands significantly. That raises very troubling questions about our predominantly national and subnational systems of governance, which historically have tended to ensure selfish local wishes take precedence over global considerations. The international agreement to phase out ozone-depleting substances is the only example of action taken to resolve a global environmental problem. In every other field, we still see leaders everywhere, whether in desperately poor countries or quite affluent ones, whether popularly elected or ruling by force, assuring their communities that further growth is not just possible but to be desired.

New technology has already helped to slow climate change, most obviously in the rapid replacement of fossil-fuel power stations by solar farms and wind turbines. We can see a clear path to electrifying much of the transport system. We need to recognise, however, that better technologies don't always prevail. When video recorders were being introduced, there were two competing systems: Beta and VHS. The Beta approach was technically superior, but VHS prevailed because it offered a wider choice of recorded material such as movies. When cars were being developed for private transport 120 years ago, the internal combustion engine was not the obvious best approach to power the vehicles, but the oil industry pushed successfully for it to be used. The other forces—pricing, regulation and community education—influence which technologies are used and how they are used. So the attitudes and consequent actions of governments play a crucial role. A striking example is the constant debate about whether nuclear power should be used and the radically different conclusions in different countries. Nuclear reactors were first built in the United States, the United Kingdom, France and Russia to provide the fissile material for nuclear weapons. Advocating their use to generate electricity was arguably a classic case of those equipped with a hammer seeing the need for power as a nail.

A CLEVER COUNTRY?

I argued earlier that we should try to become a clever country, rather than the comparatively affluent Third World nation we are today—exporting raw materials to pay for the products we are not clever enough to make for ourselves. In 1964, when I was studying part-time and earning an honest living building electronic equipment for university scientists, Donald Horne published his book *The Lucky Country*.[24] He described Australia as 'a lucky country run mainly by second-rate people who share its luck', arguing that we live mainly on other people's ideas, with leaders so out of touch with global trends that they are 'often taken by surprise'. The book was a runaway bestseller and the phrase 'the lucky country' became common usage, but the message was often misrepresented by people who had either not read the book or not noticed the irony of the title. The book sounded three warnings; reflecting on its analysis recently, I argued that they are even more relevant today than they were nearly sixty years ago.

First, Horne said, we need to recognise where we are on the map. While we have tended to engage with our neighbours purely on economic terms, we need to behave as if we plan to live permanently in this part of the world, developing foreign policies that acknowledge its complexity. He argued that we need to recognise the competing interests of the two great powers bordering the Pacific, China and the United States, as well as the concerns of middle-weight countries like ourselves and Indonesia. So we should have an independent foreign policy, rather than stupidly behaving as if we were an American colony, the role effectively taken by most recent governments. Second, Horne argued, we need to recognise that Australia is no longer the Anglo-Celtic monoculture it once was and aspire to 'a bold redefinition of what the whole place adds up to now'. We need to understand both the colonial history of dispossessing the original Australians and the complex multicultural society we have become as a result of the more recent waves of migration. Horne called for a serious public discussion of societal values, population growth and what sort of country we would like to become. Building on that point, his third proposal was that we should invest in education and science to become a clever country, in charge of our own destiny, rather than being tethered to a global economy that has absolutely no interest in our welfare.

In 2016, I revisited Horne's arguments and added the need to live within the ecological limits of this land.[25] I wrote that his proposals were, if anything, more urgent today after decades of masterly inaction by leaders responding timidly to short-term priorities, taking the easy route of allowing overseas-owned corporations to develop and export our mineral resources on terms that have done little to advance the interests of our community. I argued that we now face 'a coming global crisis, a perfect storm of limited resources, serious environmental problems, widening inequality, economic instability and political tensions'. I drew attention to the 2015 encyclical released by Pope Francis, *Laudato si'*, which argued that a sustainable future for human civilisation has to be based on two principles: social justice and a willingness to live within the limits of natural systems. Local thinker Professor Robert Manne took up the encyclical's conclusion that the present development strategy is heading for catastrophe. He warned that slowing climate change would not solve the problem if we fail to address injustice and inequality, reinforcing the papal message of the need to combine ecological responsibility with a commitment to social justice.

That raises the critical issue of poverty, which compounds the environmental problems in two fundamental ways. Most basically, in those parts of the world where many people still live in desperate circumstances, we still see high rates of infant mortality, economic instability and little chance of security in old age. In that situation, it is a rational response for a couple to have several children, in the hope that some will survive to look after their ageing parents. In those parts of the world where women are well educated, financially secure and in control of their fertility, the birthrate has declined to a level that is consistent with a stabilising of the population.

There is a second issue which is an increasing problem for those of us in relatively affluent countries. While there continue to be huge differences in material living standards between Australia and most Asian countries, between Western Europe and Africa, between North America and countries to the south, there will continue to be literally millions of people wanting to improve their lifestyle and their children's opportunities by migrating. If we don't want to see people crowding onto flimsy boats and braving shark-infested waters to try to get to Australia, we have to work to improve their living standards where they are now. I believe we should be scaling up the level of our foreign aid towards the United Nations (UN) target of 0.7 per cent of our total economic output and targeting the aid to improve the living conditions of the poorest people in the Asia-Pacific region. If we don't do that, there will continue to be an unacceptable scale of attempted economic migration, in addition to the flood of people fleeing conflict or the devastation of climate change.

What can we do to improve our chances of a decent future in Australia? In 2010, the Australian Academy of Science launched an ambitious project to consider a range of possible futures for Australia. The project was guided by two of our most outstanding interdisciplinary thinkers: Professor Tony McMichael of the ANU and Dr Michael Raupach of CSIRO. Sadly, neither of them lived to see the completion of the project. Its first stage brought thirty-five leading researchers together for a four-day workshop to explore the issues, starting from the recognition that we are now facing a new challenge: after centuries of near-continuous growth, punctuated by appalling conflicts, we need to adapt to the reality of a finite planet. The second fundamental challenge is that of increasing inequality: for the privileged, health and wellbeing have continued to improve and the rich have become steadily richer, while the poorest of the world have seen little or no improvement in their living conditions. We also live in a world that is increasingly interconnected, so events like the 2008 global financial crisis and the more recent COVID-19 pandemic ripple around the globe, affecting us all.

The report of that 2010 academy project, *Negotiating Our Future: Living Scenarios for Australia to 2050*, sets out a range of possible futures.[26] One group developed three broad scenarios, to explore the strengths and weaknesses of different approaches. The scenario they called 'Going for Growth' was broadly similar to the recent policies of our national governments, reducing regulation and promoting market-oriented solutions to emerging issues. The group concluded that this approach would allow short-term benefits of continuing economic growth, with the possibility that some of this might trickle down to the disadvantaged, but warned that recent trends showed it would be more likely to worsen the size of an underclass and the consequent social tension. The second scenario, 'Tax and Spend', involved government using higher taxes to invest in education, health and welfare. A benefit of this approach would be more equal access to essential public services, while a possible risk would be a lower rate of economic growth leading to reduced capacity for people to fulfil their other wishes. The third scenario was called 'Post-Materialism', a future in which governments set broad parameters for environmental protection and implement specific policies to limit material production and resource use as well as stabilising the population. This approach had clear social and environmental benefits, but it risked reducing capacity to adapt to changing global circumstances by forcing a lower level of economic growth.

The point of the exercise was to demonstrate that there is no obvious right answer; each possible approach has some benefits and some disadvantages. Your preference reflects your individual values—the relative importance you place on the different areas of economic, social and environmental change. My position, of course, is that since it is now abundantly clear that the present level of consumption is degrading the environment, both nationally and globally, policies to promote further growth are irresponsible—the survival of civilisation requires us to get back in balance with natural systems. And so, predictably, when fifty Australians were brought together to discuss the report—the various scenarios and the range of possible futures—there were some clear differences arising from their various backgrounds and the relative importance they saw for those three areas. What was striking, however, was what the final report of the project called a widely held preference for 'a future Australia that is more caring, community-focussed and fair than present-day Australia'.

Many of the participants, even those strongly committed to further growth of existing economic activities, were concerned about the present inequality of wealth, education, health and life opportunities. Widening inequality, they recognised, erodes social stability and weakens our ability to respond to the new challenges we don't yet recognise. There was widespread concern about what the report called 'current governance arrangements and their ability to prepare Australia for an uncertain future', with specific comments about the declining trust in governments and the lack of a shared vision guiding policy. This is a critical issue. It has been wisely said that if you don't know where you are going, any road is equally valid; if you do know where you want to go, that long-term goal informs the short-term choices of policies and programs that lead in the preferred direction.

We live in an uncertain world. There is a real probability that the overuse of natural resources and pushing natural ecological systems beyond critical tipping points will lead to serious consequences. Given the risk of disruption, we would be better prepared if we reversed the de-industrialisation of the last fifty years and embarked on a systematic program of becoming more self-sufficient, making us more resilient and better able to cope with whatever happens to global systems of trade and commerce.

THE FUNDAMENTAL ISSUE OF GROWTH

In his book *Collapse*, Jared Diamond argued that societies tend to grow and expand until they reach limits: finite supplies of water or productive land, limited mineral resources, environmental consequences of their actions, perhaps relations with neighbouring societies.[27] When this occurs, Diamond says, the society must choose whether to change its approach to manage within the limits or press bravely on, doing what it has always done. He argues that many societies kept doing what they had always done, despite the evidence they had reached limits, and collapsed. A famous example is Easter Island. Other societies, including other Pacific islands, recognised the problem and adapted to ensure their survival.

There is a striking Northern Hemisphere example. During the so-called medieval warm period, the Vikings established settlements in Iceland and also Greenland, which was at the time a green land rather than a frozen wasteland. Then the so-called Little Ice Age hit Europe, causing a significant drop in temperatures. Shakespeare wrote of milk coming 'frozen home in pail', and the Thames froze solidly enough for frost fairs to be held, with whole bullocks roasted on the ice. The settlement on the island of Greenland was no longer able to grow its food and so it collapsed. But the Icelandic community recognised what was happening and started a fishing industry, getting their protein from the sea and developing a trade in smoked fish to import the fruit and vegetables they could no longer produce. They adapted and survived.

The evidence is clear that the present level of human consumption is damaging our future prospects. We are changing the global climate, driving a loss of species, eroding the productivity of our land, reducing the fish catch and reducing the availability of such essentials as fresh water. In that context, aiming at further growth does not seem an intelligent approach. We should be prepared at least to think about alternatives. The late Professor Herman Daly came to a conference in Australia in 1977 to speak about his idea of steady-state economics. He argued that recent growth had done little or nothing to improve community wellbeing, while risking serious environmental damage.

I have followed with interest the work of the Canadian academic Professor Peter Victor, whom I met when we both gave papers to the same international conference a decade ago. He said that he grew up believing the conventional wisdom that economic growth would alleviate poverty, provide full employment and enable us to address environmental problems. But he realised that thirty years of unprecedented economic growth had not achieved any of those goals. Not only had economic

growth failed to eliminate unemployment and lessen
poverty, it had actually widened inequality and
worsened environmental problems. So he decided
to use the sort of economic models employed by the
Canadian Treasury to assess the impacts of different
approaches.[28] I was interested in his work because
Canada is very similar to Australia, as a large country
with much of its land sparsely inhabited, and using a
parallel economic approach of exporting commodi-
ties to pay for its imports of goods and services. Victor
found that a business-as-usual future in which the
traditional growth approach is used did not reduce
unemployment, and saw unacceptable environmental
damage and increasing numbers living in poverty. An
alternative approach of stopping growth immediately
was even worse, with massive unemployment and
widespread social unrest. He then analysed alternative
futures in which growth rates were slowly reduced at
the same time as other policies were introduced: a
shorter working week to spread employment more
evenly; measures to redistribute income and pollution taxes to discourage wasteful industrial practices.
He found this to be a much more attractive future,
with slower growth producing full employment,
more leisure, much less poverty and improved
environmental outcomes.

What policies were needed to produce this optimal future? First, and most fundamentally, as Victor put it, 'Managing without growth requires a stable population.' This is an obvious conclusion. If the population is growing but the economy isn't, per-capita wealth is declining. This reverses the usual wisdom from conventional economists, who like to see a growing population because this promotes economic growth: more people mean we need more food, more clothes, more houses, more transport and so on. Victor's conclusion was that we only really *need* economic growth if the population is growing. If the population grows, all other things being equal, the economy grows proportionally—but that doesn't make anyone better off. Where the population has stabilised, there is less pressure to grow the economy and in turn put increasing stresses on natural systems.

Second, and equally importantly, we need to recognise the critical importance of maintaining the integrity of those natural systems. They give us the absolute essentials of life—breathable air, potable water, the capacity to produce our food—as well as our sense of place, our sense of identity and the spiritual refreshment that comes from enjoying natural areas. Most governments see the important agencies as those which promote economic development, like Treasury,

Trade, Resources and Industry. The clear implication is that decision-makers believe environmental issues are less important, and we can always repair environmental problems if we are sufficiently wealthy. We are now seeing the flaw in this argument. No amount of wealth can restore an extinct species, or repair degraded landscapes on any realistic timescale. Even in simplistic economic terms, we are all now paying a heavy price in Australia for the damage resulting from climate change.

As Professor Victor's work showed, it is at least feasible to consider futures in which we don't regard the economy as an end in itself and see maximising economic growth as the highest priority, one which overrides human rights, social cohesion and environmental integrity. We need to change our thinking about the economy, reflect on what is important to us and others, in particular recognising the most basic flaw in the common assumption that we can safely leave the level of environmental protection to the market. There are two important groups that are critically affected by what we do to the natural environment but cannot possibly reflect their wishes in the market: all future generations, and all other species. I remember hearing an environmental advocate making the point that we don't just inherit the Earth from our parents, we also borrow it from our children. Out of common decency, they are entitled to expect us to think about the health of the country we are passing on to them. We would probably make wiser decisions in a whole range of areas if we routinely asked ourselves how our grandchildren will assess what we are doing: will they be impressed with our consideration, our vision, our wisdom and our humanity, or will they be appalled by our short-sighted selfishness?

The most obvious consequence of slowing or stopping economic growth is that it would force consideration of the distribution of wealth. We have steadily become a much less equal society in the last fifty years, as Andrew Leigh pointed out in his book *Battlers and Billionaires*.[29] If the economy is growing, each person is on average slightly better off this year than they were last year. If there were no growth and incomes were frozen at the present level, the inequalities would be so strikingly indefensible that there would have to be some remedial action. A wide range of policies adopted during the period of obsession with neoclassical economic ideology has systematically redistributed income from the poor to the rich. It is not just that professionals in such fields as medicine and law have more opportunities to reduce their taxable income than wage earners.

At the extreme, there are individuals in the list of the 500 wealthiest Australians whose declared income is so low they are exempt from the Medicare levy! It was observed in the discussion about housing, now widely seen as a medium of speculative investment rather than a basic human right, that current policies make it easier for a surgeon or barrister to buy their fifth house than a young couple to get a roof over their heads.

As a civilised society, we should be having a discussion about the policies that influence the distribution of income. Most people would say that the service workers on whom we critically depend, such as paramedics, nurses, aged-care workers, drivers of public transport vehicles and waste management operators, deserve to be better paid. There was an almost perceptible collective cheer around the country when the incoming Albanese government, in its first weeks of office, persuaded the Fair Work Commission to raise the minimum wage. While discussing the distribution of income should be part of the political discourse now, it would be forced onto the agenda if the economy were not growing.

Dr Richard Eckersley has posed a fundamental question. He asks people, 'Given what you know about the state of the world and your own financial situation, is it absolutely your highest priority to become twice as wealthy in the next twenty years and consume twice as much?' He says that very few people respond in the affirmative. Some don't even see that scale of increase in wealth and consumption as desirable. Those who do see it as desirable still usually rank it below other goals such as staying healthy, having satisfying work, being in a strong relationship, feeling secure in their neighbourhood and seeing children happy. But all governments, state or Commonwealth, Coalition or ALP, seem to believe that the absolute highest priority is to ensure that the gross domestic product, the total size of the economy, grows at a rate of at least 3.5 per cent—in other words, to double the economy in twenty years. That would only be sensible if we all agreed that becoming twice as wealthy in the next two decades is our principal objective, justifying widening social divisions, accelerating environmental damage and increasing foreign control of our productive land.

In principle, we could decide to curb consumption and reduce the rate of economic growth almost immediately. What we cannot change in the short term is the rate of population growth. Even such radical moves as China's one-child policy take decades to have a significant impact on population numbers.

A DEMANDING POPULATION

The issue of population growth is one of the most contentious questions in political circles. Part of the problem is the spread of misconceptions, which perhaps we should now call 'fake news'. It was reported that the birthrate in Australia in 2022 was about 1.6 children per adult woman. This has changed dramatically in my lifetime, from the postwar period when the average woman had about four children. Women are now better educated, more able to control their fertility and more likely to be interested in careers outside the home. Because the birthrate is now less than the replacement rate, I have seen uninformed comment that the population would be shrinking if we weren't bringing in migrants. When migration was effectively halted by the COVID-19 pandemic, there was mild panic in some areas of business, so the Albanese government has been strongly urged to restore migration levels to those we had before 2020.

What the simplistic analysis ignores is that there are two factors determining how many babies are born: the number of children per adult woman and the number of adult women. In fact, the number of women in the reproductive age ranges is still increasing as a result of the past birthrate and earlier waves of migration. The critical number is the so-called 'natural increase': the difference between the number of births and the number of deaths. That has been well over 100 000 a year for as long as I can remember. It was about 120 000 a year when the Howard government claimed it was a problem and brought in a 'baby bonus' to encourage women to have more children. The then treasurer, Peter Costello, fatuously called on couples to have three children: 'One for Mum, one for Dad and one for the country'—as if it was almost their duty to accelerate our population increase. Whether it was a result of the bonus or the windy rhetoric is unclear, but the natural increase jumped to about 150 000 a year. So the situation before the pandemic was that the Australian population would have been increasing by 150 000 a year, or a million people about every seven years, if there had been no migration. In fact, the net migration was typically about 200 000 a year, meaning the population was increasing by about a million every three years.[30]

What will happen in the future? As I have already asserted, there is not one predetermined future but a range of possible futures. If there were no net migration from now on, the natural increase would gradually shrink as the population ages, and the population would stabilise some time about 2040. At net migrant intakes up to about 70 000 a year, the population stabilises later at a higher level. For net migrant intakes above 70 000 a year, the population continues to increase for the foreseeable future. In practical terms, about 80 000 people left Australia in a typical year before COVID intervened. So any immigration levels up to about 150 000 a year are consistent with a long-term goal of stabilising the population.

To be clear about my position, we now know that the demands of the current Australian population are degrading our natural systems. Increasing the population without fundamental changes in lifestyle will increase proportionally the damage we are doing. So I believe it would be responsible to have a long-term goal of steadying our population. That will take at least fifteen to twenty years, during which time our impacts on the natural systems of Australia will continue to worsen the situation. If we want to get back into balance with those systems, we need both to slow the rate of population growth and systematically reduce our average per-capita demands. We could still be more generous to refugees, but we should be less welcoming to cashed-up entrepreneurs.

As the first report to the Club of Rome pointed out fifty years ago, the most fundamental problem is the naive assumption that there are no limits to growth. Both nationally and globally, we need to recognise there are limits and that current consumption is pushing us dangerously past some of those limits. It is no exaggeration to say that the survival of civilisation requires us to work together to bring total human consumption back to a level that maintains the integrity of natural systems.

A SUSTAINABLE FUTURE AUSTRALIA?

When I wrote *A Big Fix*, I set out what I saw as the criteria for a future sustainable Australia. I made the point that aiming for a sustainable future is nothing more than our moral duty, because developing in a way that is not sustainable is stealing from our own children. This is an updated version of my basic checklist for that future:

1   It will have achieved genuine reconciliation with the First Australians.
2   It will have stabilised the human population.
3   It will be using renewable resources within the rate at which they are produced by natural systems.
4   It will be using non-renewable mineral resources at a rate that allows development of replacements.

5   It will be approaching the goal of zero waste, committed to recycling and reuse of resources.

6   As well as being committed to maintaining the natural areas that still exist, it will be restoring critical assets that have been degraded by earlier misuse.

7   It will have serious environmental standards for new developments, with an Environmental Protection Agency resourced to enforce those standards.

8   It will be a low-carbon society, meeting its energy needs from a mix of renewable energy resources, principally solar and wind, with storage to smooth out variations.

9   It will be much more equitable than today's society, reducing the pressure to expand consumption levels.

10 It will have a more mature politics with an inclusive process for making difficult decisions, recognising that change will always have losers as well as winners, and accepting the principle that those who lose for the common good should be compensated in some way.

That is, of course, very different from the Australia we live in today, and you may think it a bit utopian. It is, but almost all of the significant changes throughout human history were seen as utopian when they were first proposed. The abolitionists were told the economy could not function without slave labour. The suffragettes were told society would crumble if women were allowed to vote. Yes, that is ancient history, but only thirty-five years ago it was utopian to be dreaming of Berlin without the Wall, or South Africa without apartheid, or female political leaders in Australia, or an African American elected US president, or Australians voting overwhelmingly to accept same-sex marriage. I sometimes joke that thirty-five years ago it was still utopian to be hoping for good coffee and civilised licensing laws in Queensland! The basic point is that change happens when enough people want it badly enough and are prepared to work purposefully for it. Now that we know how much damage we have been doing to our environment locally, nationally and globally, we should all feel obliged to be working to get us back into balance.

Avoiding environmental ruin and the potential collapse of civilisation requires significant cultural change. Dr Paul Raskin, who heads the Boston-based Tellus Institute, argues that the dominant values in countries such as Australia and the United States are individualism, consumerism and domination of nature.[31] These may have served us well in the past and resulted in material living standards being higher than ever before, but they are now obstacles to the fundamental changes we need to make. Individualism needs to be replaced by a recognition that we are all in this together, facing global problems that demand global solutions. Our collective survival requires social justice as well as a willingness to live within the limits of natural systems. Consumerism, the endless pursuit of more goods and services, needs to give way to a commitment to optimising the human experience; as a former colleague said to me, those who win the rat race are still rats! Finally, domination of nature needs to be replaced by what could be called ecological sensitivity, recognising that natural systems have critical limits and accepting the need to live within those limits. Just as the protesters say there are no jobs on a dead planet, there is no music on a dead planet, no literature or art, none of the leisure pursuits that make our lives worthwhile. Our civilisation can only survive if we maintain the integrity of the natural systems of this small blue globe. This is the critical decade, what the UN called the rapidly closing window of opportunity. We owe it to our children and grandchildren to be working purposefully for a future that is, at least in principle, sustainable.

What is preventing that change? A discussion at the 2022–23 Woodford Folk Festival produced a strong feeling that the single factor doing most to prevent progress is allowing corporate donations to political parties. It would be naive to believe that corporations donate large sums to political parties out of philanthropy. Their shareholders would probably be angry if this were the case! Corporations expect that politicians will be reluctant to bite the hand that feeds them. Large fossil-fuel companies have often been even-handed, donating both to the Coalition and the ALP, so that whoever is in government will feel kindly towards them. If election campaigns were funded from the public purse, with parties receiving sums reflecting their support levels in previous years, there would be less temptation for governments to look generously on corporate donors.

I was briefly funded by the Tasmanian Government more than thirty years ago to produce an alternative development plan based on the principles of living sustainably. I called the alternative blueprint a 'greenprint'. Interestingly, it showed that developing in ways that used natural resources sustainably was just as good economically as the traditional approach, as well as being better socially and much better environmentally. A similar exercise was being conducted at the same time in South Australia, leading to the same broad conclusions in the report *Our Future State*. Organisations like the SA-based Wakefield Futures Group and the Australian Earth Laws Alliance have been working to develop these greenprints—plans for more responsible development strategies.

In an ideal world, government agencies would be putting serious resources into that sort of work. Unless we have a clear vision of a better way and a viable pathway to get there, elected politicians will continue to dither and stick to what they have done previously, even if they are aware of the looming problems. Now that we know we are on the ecological *Titanic* and heading for an iceberg, we need to put serious effort into steering towards clear water. Like the *Titanic*, societies have a lot of inertia and take time to change direction, but the situation is now urgent.

GROUNDS FOR CAUTIOUS OPTIMISM

I was heartened to read about a 2022 exercise in South Australia. Professor Andy Lowe (no relation) led an ambitious exercise, bringing 150 people together in the rural setting of the Ukaria Cultural Centre to discuss the future of their state. From the twelve discussion groups, the report of the process notes that three common themes emerged. They are worth quoting:

- unsustainable growth: of population, economies and consumption
- widespread loss of connection between people and place
- an urgent need for adaptive systems of governance.[32]

The report concluded that the challenge we now face is to recognise those issues and work towards 'a dynamic, transitional state'. With a local emphasis, it said that the key goal now is to 'inspire the SA community to actively transform our State and connect deeply with people and place'.

Of course, neither those themes nor the suggested way forward are specific to South Australia. While that state has a long tradition of progressive thinking, both the issues identified and the proposed path forward apply to Australia more generally. Faced with unsustainable growth of population, economies and consumption, as well as widespread loss of the historic connection between people and the places where they live, we urgently need adaptive systems of governance to transform our nation, redirecting the trajectory of development, and restoring our connection with this ancient land.

The basis for my cautious optimism is the recognition that human systems are non-linear and can change very rapidly from one stable state to another. In six months in 2007, John Howard went from being unassailable to being unelectable, from one of our longest-serving prime ministers to losing his own electorate. The 2022 election saw independents and Greens elected in seats that had historically always voted for one of the major parties. Again, change happens when enough people want it badly enough and work purposefully to achieve it. Our situation is now critical. There is no time to lose. I hope you will do what you can to help redirect our path of development.

ACKNOWLEDGEMENTS

I am grateful to Monash University Publishing for the invitation from Greg Bain to contribute to this series. The text was significantly improved by perceptive comments on the first draft from Professor Rob Fowler and Peter Martin, my colleagues in the Wakefield Futures Group. I also acknowledge the improvements which resulted from the sensitive editing by Paul Smitz. The arguments in this book have been refined by discussion with many generations of students. The errors and oversights that remain are entirely my responsibility.

NOTES

1   I Lowe, *A Big Fix: Radical Solutions for Australia's Environmental Crisis*, Black Inc., Melbourne, 2005. An updated version was published in 2009.
2   State of the Environment Advisory Council, *State of the Environment Australia*, CSIRO Publishing, Collingwood, Vic., 1996.
3   Council of Australian Governments, *National Strategy for Ecologically Sustainable Development*, 1992, https://documents.parliament.qld.gov.au/tp/2016/5516T2036.pdf (viewed February 2023).
4   T Plibersek, 'Labor's Nature Positive Plan: Better for the Environment, Better for Business', media statement, Department of Climate Change, Energy, the Environment and Water, 8 December 2022, https://minister.dcceew.gov.au/plibersek/media-releases/media-statement-labors-nature-positive-plan-better-environment-better-business (viewed February 2023).
5   SG Kearney et al., 'The Threats to Australia's Imperilled Species and Implications for a National Conservation Response', *Pacific Conservation Biology*, vol. 25, no. 3, April 2018, pp. 231–44, https://www.publish.csiro.au/pc/fulltext/pc18024 (viewed February 2023).
6   Australian Government, *Australia: State of the Environment 2021*, 2021, https://soe.dcceew.gov.au/ (viewed February 2023).

7   Bureau of Meteorology and CSIRO, *State of the Climate 2022*, http://www.bom.gov.au/state-of-the-climate/2022/documents/2022-state-of-the-climate-web.pdf (viewed February 2023).

8   Climate Council of Australia, *The Great Deluge: Australia's New Era of Unnatural Disasters*, 28 November 2022, https://www.climatecouncil.org.au/resources/the-great-deluge-australias-new-era-of-unnatural-disasters/ (viewed February 2023).

9   Intergovernmental Panel on Climate Change, *Summary for Policymakers*, 2022, https://www.ipcc.ch/report/ar6/wg2/downloads/report/IPCC_AR6_WGII_SummaryForPolicymakers.pdf (viewed February 2023).

10 United Nations Environment Programme, *GEO-5: Environment for the Future We Want*, 2012, https://www.unep.org/resources/global-environment-outlook-5 (viewed February 2023).

11 WJ Ripple et al., 'World Scientists' Warning to Humanity: Second Notice', *BioScience*, vol. 67, no. 12, 2017, pp. 1026–8, https://doi.org/10.1093/biosci/bix125 (viewed February 2023).

12 WWF, *Living Planet Report 2022*, https://livingplanet.panda.org/ (viewed February 2023).

13 United Nations, *Resource Efficiency—Economics and Outlook for Asia and the Pacific*, 2011, https://sdgs.un.org/publications/resource-efficiency-economics-and-outlook-asia-and-pacific-17168 (viewed February 2023).

14 DH Meadows et al., *The Limits to Growth*, Universe Books, New York, 1972.

15 A Finkel, 'Getting to Zero: Australia's Energy Transition', *Quarterly Essay*, no. 81, Black Inc., Melbourne, 2021.

16 A Blakers, 'Extra 1500 Pumped Hydro Sites Offer Potential to Boost Australia's Clean Energy Grids', Australian National University, 11 November 2022, https://re100.eng.anu.edu.au/2022/11/11/extra-1500-pumped-hydro-sites-offer-potential-to-bolster-Australias-clean-energy-grids/ (viewed February 2023).

17 I Lowe, *Long Half-Life: The Nuclear Industry in Australia*, Monash University Publishing, Clayton, Vic., 2021.

18 B Brook and I Lowe, *Why vs Why Nuclear Power*, Pantera Press, Sydney, 2010.

19 Australian Institute of Refrigeration, Air conditioning and Heating, *Towards a National Framework for Energy Efficiency*, Melbourne, 2003, http://airahfiles.org.au/Advocacy/Sustainability/NFEE_Stage1_DiscussionPaper-2004.pdf (viewed February 2023).

20 F Castro-Alvarez et al., *The 2018 International Energy Efficiency Scorecard*, American Council for an Energy Efficient Economy, Washington, DC, June 2018.

21 PWG Newman and J Kenworthy, *Sustainability and Cities: Overcoming Automobile Dependence*, Island Press, Washington, DC, 1999.

22 S Griffith, *The Big Switch*, Black Inc., Melbourne, 2022.

23 Clive Hamilton and Hal Turton, 'Subsidies to the Aluminium Industry and Climate Change', Background Paper No. 21, The Australia Institute, Canberra, 1999.

24 D Horne, *The Lucky Country*, Penguin Books, Melbourne, 1964.

25 I Lowe, *The Lucky Country? Reinventing Australia*, University of Queensland Press, St Lucia, 2016.

26 M Raupach et al., *Negotiating Our future: Living scenarios for Australia to 2050*, Australian Academy of Science, Canberra, 2012.

27 J Diamond, *Collapse: How Societies Choose to Fail or Succeed*, Penguin Books, Harmondsworth, 2005.

28 P Victor, *Managing without Growth: Slower By Design, Not Disaster*, Edward Elgar, Cheltenham, UK, 2008.

29 A Leigh, *Battlers and Billionaires*, Black Inc., Melbourne, 2018.

30 I Lowe, *Bigger or Better? Australia's Population Debate*, University of Queensland Press, St Lucia, 2012.

31 P Raskin, *Journey to Earthland: The Great Transition to Planetary Civilization*, Tellus Institute, Boston, 2016.

32 AJ Lowe et al., *Dynamic Statement: Towards a Regenerative Future for South Australia*, 2022, https://dynamicstate.com.au/wp-content/uploads/2022/12/DYNAMIC-STATEMENT-2022.pdf (viewed February 2023).